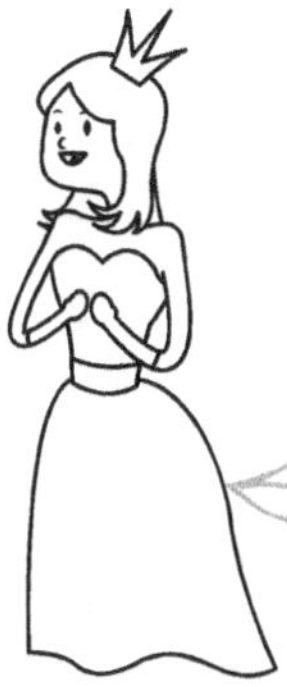
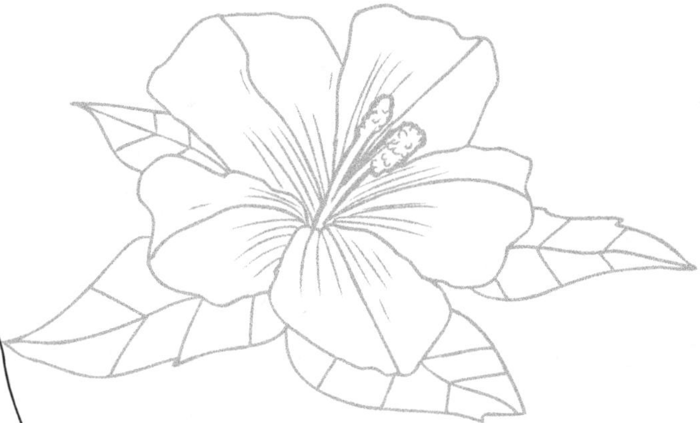
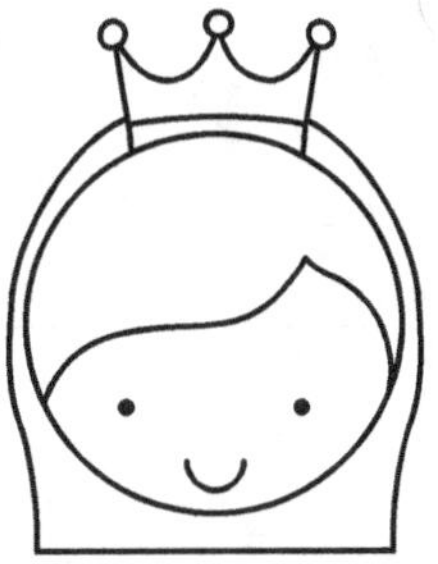

THIS BOOK BELONGS TO:

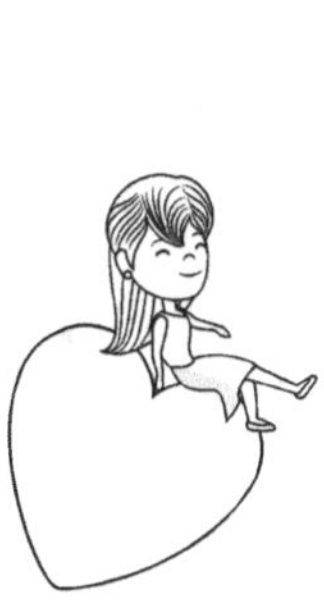
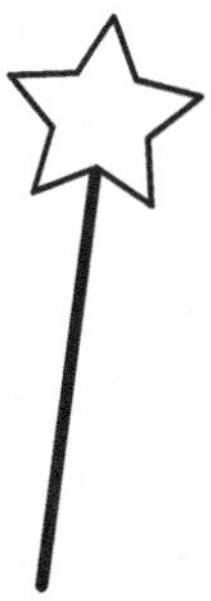

www.ingramcontent.com/pod-product-compliance
Lightning Source LLC
Chambersburg PA
CBHW081430220526
45466CB00008B/2336
9781655222696